TRAINI
PIS

"From Safety To Skill: A Novice's Journey Into Pistol Mastery"

JOSHUA ANDREW

Contents

CHAPTER 1

INTRODUCTION AND BASICS

Pistols are exceptional weapons for both self-defense and recreation. However, it is a tremendous responsibility to handle one. In the end, a handgun does not qualify as a play.

Acquiring the necessary skills to fire a pistol accurately is critical for ensuring gun safety and equipping oneself to defend one's family in the event of an emergency. We will examine some of the fundamentals of handling and firing a handgun correctly in this article. Further reading will instruct you on the correct way to fire a handgun. Ensure that your handgun is unloaded (no loaded magazine is inserted and the chamber is clear) prior to attempting any of the below. Put safety first, then enjoy yourself.

FIREARM SAFETY GUIDELINES

Safety first at all times, whether handling a firearm at the range or elsewhere. Here are the four fundamental firearms safety rules.

By emphasizing the fundamentals of secure firearm storage and safe handling and by reminding you that you are the key to firearms safety, this list is intended to emphasize these points. When handling firearms, you must consistently emphasize safety, particularly to minors and non-shooters. Particularly vulnerable individuals must be under close supervision when interacting with armaments that they are unfamiliar with.

1. Keep the firearm pointed in a safe direction at all times. Avoid aiming your firearm at any object that you do not intend to strike. This holds specific significance in the process of loading or discharging a firearm. As long as the muzzle is aimed in a secure direction, accidental discharge can be avoided without causing any harm. A safe trajectory is one in which it is implausible for a bullet to strike an individual, considering factors such as ricochets and the ability of bullets to penetrate walls and ceilings. While the safe trajectory may fluctuate between "up" and "down" at times, it is never directed at an object that is not intended as a target. Even when "dry firing" with an unloaded firearm, a hazardous target should never be in the line of sight. Constrain yourself to always be aware of the precise direction in which the muzzle of your firearm is pointing, and ensure that you maintain control over that direction notwithstanding any potential missteps or falls. It is solely your responsibility to exercise control over this.

2: Consider every firearm to be loaded

By approaching each firearm as if it were loaded, one cultivates a safety-conscious routine. Loading firearms should only occur when one is prepared to fire from the target range, shooting location, or field. Before handling or transferring a firearm, you must at all times open the action and visually inspect the chamber, receiver, and magazine to ensure they do not contain any ammunition. When not in use, actions should always remain accessible. It is never safe to presume a firearm is unloaded; always verify it yourself. Such behavior is indicative of a proficient firearms operator.

3. Maintain your hand away from the trigger until you are prepared to fire

Touch the trigger of a firearm only when you have every intention of pulling the trigger. It is advised to prevent finger contact with the trigger when loading or discharging. It is strictly forbidden to depress the trigger of a firearm while the safety is in the "safe" position or any location between "safe" and "fire." There exists a potentiality for the firearm to discharge at any moment, or even subsequent to the release of the safety, in the absence of subsequent trigger manipulation.

4. Constantly Be Certain of Your Objective and the Area Beyond It

Avoid shooting until you have complete knowledge of the target that your bullet will strike. Ensure that nothing or no one beyond the intended target will be injured by your projectile. It is worth noting that a.22 short projectile has the capability to travel over 1 1/4 miles, while a.30-06 high velocity cartridge can propel its bullet over three miles. Shotgun projectiles have a range of 500 yards, while shotgun slugs can travel over a half-mile. It is advisable to consider the distance a projectile will traverse in the event that it deviates from its intended target or undergoes ricochet.

Pistol Parts

Whether you are an aspiring shooter, a neophyte gun enthusiast, or simply someone interested in firearms, it is vital that you have a fundamental understanding of the components of a handgun. Understanding how this assembly of components functions to create a functional firearm is crucial for the responsible and secure operation of firearms. We shall examine the fundamental constituents of a handgun, as well as the evolutionary trajectory of handguns and their operational components.

1. Muzzle and Barrel: A handgun's barrel is an essential component of its construction. It is the metallic conduit through which the discharged projectile proceeds. The muzzle is the area at the front of the barrel through which the projectile emerges. The polish and design of the muzzle can have a substantial impact on the bullet's velocity and accuracy. Additionally, some handguns provide the capability of a suppressor being affixed to the muzzle in order to diminish the flash and commotion generated while discharging. The barrel not only directs the projectile towards its intended target, but its length and internal construction also have a substantial influence on the bullet's velocity and precision. The lumen of the barrel, referred to as the bore, and the existence of rifling, which consists of helical grooves encircling the barrel, both impact the bullet's muzzle velocity and stability during flight.

2. Trigger and Trigger Guard: The trigger, which the shooter activates by pulling a lever, is the focal point of the discharge sequence. The trigger guard is a protective coil that encircles the trigger with the purpose of averting inadvertent discharge. Due to the critical nature of responsible trigger use, firearm handling safety is compromised; therefore, the trigger guard is crucial for overall firearm safety.

3. Firing Mechanism and Firing Pin: The operational core of a handgun is its firing mechanism. Pulling the trigger initiates the firing mechanism, which results in the discharge of the firing pin. The powder within the cartridge is ignited and the projectile is propelled forward when the cartridge primer is struck by this pin. Firing mechanism accuracy is critical for dependable firearm operation.

4. Rear Sight and Front Sight: Aiming the handgun is facilitated by the rear and front sights. A post typically comprises the front sight, whereas the rear sight is a diminutive depression. By aligning the post with the depression, the handgun can be precisely aimed at the target.

5. Pistol Grip: The interface between the gunman and the firearm is the pistol grip of a handgun. It is ergonomically designed to be held in the hand and absorbs recoil from the firearm while the shooter is operating it. A pistol grip that is ergonomically designed can enhance both shooting precision and comfort.

6. Slide and Slide Lock: The slide, which is the upper section of a semi-automatic handgun, retracts the expended casing and enables the reloading of a new round from the magazine when the firearm is discharged. After the final round has been discharged, the slide lock maintains an exposed slide, serving as a visual indication that the handgun is devoid of any ammunition.

7. Safety Mechanism: Pertaining to the prevention of inadvertent discharges, the safety mechanism of a handgun is an indispensable component. It may manifest in diverse modalities, including a lever or button, and its operational purpose is to obstruct the trigger or discharge mechanism upon activation.

8. Decocking Lever and Takedown Lever: Semi-automatic pistols frequently incorporate these features. The decocking lever enables the hammer to be descended in a secure manner, preventing the discharge of a round. Conversely, the takedown lever streamlines the process of disassembling the firearm to perform maintenance and cleaning duties.

9. Recoil Spring: Contributing to the slide's forward motion in preparation for the next round, the recoil spring absorbs the force of the slide's backward movement after a round is discharged. This spring is vital to the operation of the cycle of a semi-automatic handgun.

10. Magazine: A removable component present in semi-automatic pistols, the magazine serves the purpose of storing and reloading ammunition into the weapon. The amount of ammunition that a magazine can contain is dependent on its design. It is inserted into the handgun's grip, and after the previous round has been discharged, a spring mechanism propels each round into the chamber.

CHAPTER 2

TO GET STARTED

Picking the Right Pistol

The police, military, and concerned citizens must select a handgun that meets their requirements in an objective and exhaustive manner. As the list of handguns is extensive and there is no ideal handgun, caliber, or projectile, making the decision challenging. It is more essential to purchase a firearm with which you are comfortable shooting rather than one that you believe you require. Beyond the "right" caliber or projectile, comfort in gun handling and firing is of paramount importance.

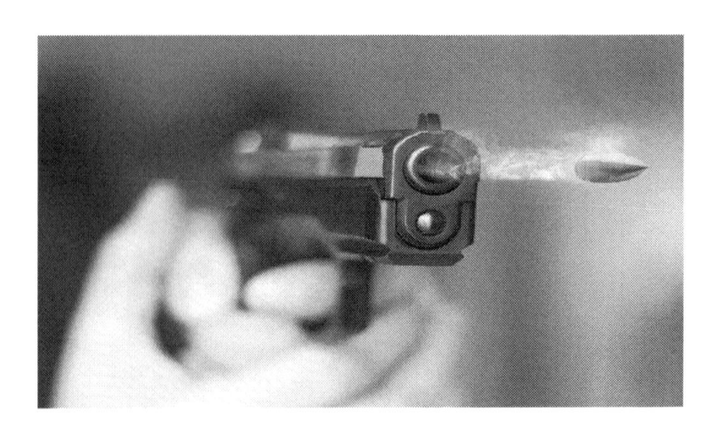

1: Consider your personal defense requirements first.

The almost exclusive use of pistols is in self-defense. Therefore, you must consider the reason you require the pistol and the locations where you intend to carry it. Do you intend to always transport it with you, or will it remain in the house? Do you desire something that will deter criminal activity and stop people in their tracks, or do you prefer something with sufficient force to violently end the standoff?

2: Figure out your basic shooting range before you buy.

You must determine whether or not you can aim a firearm with ease prior to purchasing it. Close your eyes and aim an unloaded firearm at an improvised target while directing the weapon in a safe direction with your finger placed near but not on the trigger. Upon opening one's eyes, the images should be precisely aligned as intended. The object should be within a couple of centimeters of center-target at a distance of five yards.

3. For improved accuracy and range, select a larger pistol.

Larger pistols are more precise than their smaller counterparts by virtue of their extended sight plane, weight (which reduces felt recoil), and ergonomics. However, they are considerably more difficult to conceal and transport, and for some individuals, the added weight may impede their ability to aim and maneuver.

4. Select a pistol with a reduced size if you must transport it on your person. While smaller pistols are lighter and simpler to conceal, they sacrifice some accuracy and power. However, due to their diminutive size, they are frequently simpler to target, particularly for smaller individuals.

STANCE OF PISTOL SHOOTING

The firing posture of an individual is an essential, albeit frequently disregarded, foundational aspect in the process of discharge from a firearm. Although there is no universally applicable "correct" shooting posture, various stances can be considered optimal, contingent upon individual preferences and level of expertise. As a law enforcement officer, it is critical to understand that your shot will be inconsistent regardless of your knowledge or skill level (including grip, sight, and recoil control, among others). Your stance must provide a solid foundation for your firing.

Stance Of Isosceles Shooting

The isosceles posture derives its name from the isosceles triangle, which consists of two sides of equal length. The base of the triangle represents the shooter's torso, while the two sides represent the shooter's arms. With the toes pointed at the target, the feet should be spaced shoulder-width apart (or slightly broader) in an isosceles stance. A forward lean and a minor flexion of the knees are typical while maintaining this stance. It is necessary to extend the limbs in the shape of an isosceles triangle.

Stance Of Weaver Shooting

Numerous new shooters are instructed in the weaver shooting posture, which was frequently attributed to Jack Weaver, sheriff of Los Angeles County in the 1950s. In order to acquire the stance, one must advance their non-dominant leg (typically the left) in front of their dominant leg (typically the right). The arm and hand that will draw the trigger, the firing arm, are completely extended, whereas the supporting arm is slightly bowed. The shooter will once more direct their toes in the direction of the target while leaning slightly forward.

Stance Of Tactical Shooting

A popular law enforcement posture, the tactical stance, is also referred to as modified or modern isosceles. It evolved from the weaver and isosceles shooting stance. The shooter assumes a square stance to the target, spaced at least shoulder width apart (or slightly broader) with their feet. The support side foot is positioned just barely behind the firing side foot. (a "middle ground" in footwear positioning between the weaver and isosceles stances). The shooter maintains an elongated forward lean and a minor knee flexion while extending their arms in a straight line.

CHAPTER 3

FUNDAMENTALS OF SHOOTING

CONTROL THE TRIGGER

Trigger control is the process of activating a firearm by manipulating the trigger while maintaining sight. The manner in which one pulls the trigger of any firearm can determine the success or failure of a given projectile. Without a strong trigger draw, having the best grip, the best stance, and the best firearms will be in vain, even if you have your sights set on the target. Mastering the "perfect trigger pull" requires extensive knowledge and practice of numerous minute details. Let's begin by discussing the rationale for appropriate trigger control.

Proper trigger control serves a singular purpose, which is to discharge the firearm while minimizing any discernible disruption to the sight alignment and sight picture.

The manner in which you release the trigger may be contingent on your "mission." When a target is in close proximity, quickness becomes crucial. We are able to tolerate greater disturbance of our sights, and our trigger draw does not require extreme precision. When engaging a target at a distance, say 15 yards or more, however, precise trigger control is required to ensure minimal eye disturbance.

Principles of Shooting

Assemble the firing principles required to achieve the intended function of the trigger control. While I may have reservations about some techniques, I strongly advocate for the observance of sound principles. A great deal of industry participation is devoted to techniques to the point where ceaseless debates ensue regarding which technique is superior.

To tell the truth...

Technique is all about style.

Diverse factors, including but not limited to body mass, musculature, and firearm magnitude, contribute to the minor variation in techniques that set each individual apart. The fundamental argument is that while techniques may vary, principles remain consistent across all shooters. Leverage remains leverage, while friction and gravity remain friction. Two fundamental principles must be adhered to, irrespective of technique, in order to achieve optimal trigger control.

By adhering to these principles, the technique one selects becomes largely inconsequential.

1: Distinguishing the Trigger Finger's Action

The initial principle is straightforward: when breaking a shot, only your trigger finger should move. Simple once more, but not effortless. To illustrate the concept of "moving only your trigger finger," consider the following analogy: while extending your arm as if you were firing, move your trigger finger as if you were commanding someone to "come here." You must take care not to motion or squeeze the remaining digits while performing this action.

You should ideally only flex your finger at the second joint. In order to increase velocity, one will inevitably initiate motion at the third joint, specifically the knuckle, due to the nearly impossible task of isolating it while executing rapid shots. However, shooting quicker typically indicates a much larger or closer target, so isolating the trigger finger can be compromised in exchange for increased velocity. To improve your trigger finger isolation, simulate a shooting grip by relaxing all of your digits while using your dominant hand. Then, gently extend your trigger finger and initiate a back-and-forth motion beginning at the second knuckle. As you gradually increase the speed, maintain a vigilant observation of your other digits. Any sympathetic movement of the other digits should be avoided.

2: Directly to the rear

The second principle pertains to the trajectory of the trigger press, specifically in the direction of rearward motion. Some of you may be perplexed as to how this is feasible given that our fingertips form an arch. Pulling a trigger directly back from a completely prepared trigger is crucial; otherwise, the trigger may break the shot if slapping or employing the zipper pull technique (further elaborated upon in the trigger manipulation section). You need only be concerned with pulling your trigger back immediately from a completely prepared trigger or as you approach the point at which it breaks. Following this, maintaining a flat finger on the trigger face will facilitate a straight backward draw.

In general, trigger control is mental control. While certain trigger manipulations require physical execution (which I will describe in greater detail below), trigger control is wholly mental.

Minimal Effort and Constant Speed

The proper implementation of the two principles I discussed is supported by two concepts. Concepts are equally as crucial to proper trigger control as principles. The initial notion concerns the velocity of the physical printing press. Again, we are discussing the actual press required to break the projectile; in this case, it is irrelevant how quickly you prepare your trigger or remove slack. Staying at a constant pace is crucial. You should complete your trigger stroke at the same speed you began it at, which is 4 mph. Numerous individuals become 'charged.' They might begin with a trigger pull of 4 mph but ultimately shatter the shot with a pull of 25 mph. By doing so, shooters produce an impulse, such as a slap or jolt, which has the potential to impact accuracy. The ideal trigger draw resembles the piston of a vehicle's engine in that it occurs at a constant, smooth speed.

The second concept pertains to the exertion required to complete a break shot.

What level of exertion is required to depress a trigger?

The answer is straightforward—exactly so.

A 6-pound trigger requires a minimum of 6.01 pounds of pressure in order to rupture the projectile. While it is not possible to provide an exact weight requirement, the overall point is that firing would require only a marginally higher pressure than six pounds. Numerous shooters exert excessive effort. For instance, their pull may be 25 pounds when 6 pounds is all that is required.

What do you believe the destination of the surplus energy is?

The energy will be transferred to the gun's frame, potentially disrupting the accurate alignment of the sights and leading to an imprecise discharge.

The most difficult aspect is maintaining a firm and rigid grasp while performing trigger manipulations with minimal force—similar to how one types on a keyboard or mouse.

By combining consistent velocity with minimal exertion, you can effectively separate your trigger finger and improve your straight-back draw.

Stages of Trigger Pull

Control and trigger discipline are predominantly mental processes.

Before I delve into the technical implementation and physical manipulation of a trigger through various trigger manipulations, I would like to provide a concise overview of its phases to ensure that you are familiar with the terminology I employ.

Trigger pull is comprised of four main phases:

Pre-travel: The motion of the trigger from its initial beginning position to the point where it is moved, which initiates the discharge of a firearm. As components of pretravel, the terms take-up (also known as trigger laxity), the wall, and creep are included. Any 'positive' motion of the trigger that does not elicit a response from the sear or engage the mainspring is referred to as "take-up." The initial point of contact between the trigger action and the sear resistance is at the wall. Any 'positive' motion of the trigger that induces motion in the sear and engages the mainspring is referred to as claw.

Break: The moment the sear releases the hammer (or striker, depending on the sort of action) during the trigger action. At this point, the cannon discharges with a loud bang.

Over-travel: Any "positive" movement following the break in the trigger.

Reset: When the trigger executes a 'negative' motion away from the user, the sear (or striker, depending on the action) re-engages, allowing the weapon to be discharged once more.

Vision Alignment

The shorter distance between the sights in pistol shooting exacerbates the significance of sight alignment, which is critical when launching a rifle.

• Handgun sights generally comprise a weighty square front blade sight and a square rear notch sight. This configuration can be easily aligned.

• The sight-in distance for the majority of handguns is 50 feet.

The objective

• While practicing hand gunnery at the range, a considerable number of individuals employ a sight picture in which the bull's-eye is positioned atop the front sight as opposed to above the target's center. Hunters must maintain the alignment precisely over the critical area.

• Long eye relief scopes have gained popularity among hand gunners and provide hunters with precise sighting. Although they require more time to align with a target than open sights, scopes are typically more precise.

While aiming your handgun, adhere to the following guidelines.

• Concentrate on the front sight when employing an open sight. The rear sight and the target should appear hazy or blurry.

• For novices, target the lower center of the target, which is the bull's-eye. This is referred to as the "six o'clock hold" because the location on the face of an analog clock corresponds to six o'clock.

• While maintaining both eyes open, aim with your dominant eye. This will improve your depth perception and illumination.

• Understand that while aiming, you cannot hold the handgun absolutely still. To reduce movement, take breaks between rounds and maintain a loose grip on the firearm.

Sight Picture:

In order to achieve accurate aim, a shooter must possess the proper sight picture. When the sights are in alignment with the target, the resulting image is known as the "sight picture." This is the case regardless of the sort of sight utilized while carrying a handgun, rifle, or other firearm. Sight photography is not dependent on innate shooting ability. Conversely, the shooter centers objects prior to delivering a projectile by employing a precise system.

CHAPTER 4

TROUBLE SHOOTING

SHOOTING COMMON MISTAKES

Target shooting is a demanding and gratifying activity that necessitates accuracy, concentration, and self-control. Whether you're a beginner or an experienced shooter, it's critical to understand the most common mistakes that might impair your accuracy and prevent you from improving your shooting abilities. We will examine the most frequent errors committed when shooting targets with pistols and rifles, along with advice on how to prevent them.

Poor Stance And Grip

An essential component in achieving precise firing is maintaining an appropriate grip and stance. The significance of maintaining a consistent and secure grasp on the pistol or rifle is often disregarded by numerous shooters. A weapon that moves during the discharge of a loose grip may result in inconsistent projectiles. In the same way, an improper posture can impair your stability and balance, leading to a lack of confidence in your aim. Maintain a relaxed, supportive grasp with your non-dominant hand while maintaining a firm grip with your dominant hand. In addition, ensure that your body is aligned with the target and that you maintain a balanced and stable shooting posture, with your feet shoulder-width apart.

Improper Picture And Sight Alignment

Proper sight alignment and sight picture are indispensable for ensuring precise firing. Sight picture refers to the alignment of the sights with the target, whereas sight alignment concerns the front and rear sights of the firearm. The most frequent error is neglecting the alignment with the rear sight and the target in favor of concentrating solely on the front sight. It is essential to ensure that the front and rear sights are aligned in order to produce a distinct

and focused sight image of the target at the same time. Improve your accuracy by adhering to proper sight alignment and picture technique.

The Trigger Jerk

The trigger pull is frequently the cause of imprecise shots. When the trigger is jerked or excessive force is applied too rapidly, the firearm may deviate from its intended target during the act of firing. It is critical to maintain a controlled, fluid trigger draw so as not to interfere with your target.

Put into practice a deliberate and consistent trigger squeeze, maintaining an even pressure until the projectile spontaneously breaks. This method reduces any disruption to your sight alignment and contributes to enhanced precision.

Planning for the Shot

Commonly committed error, anticipating the shot may lead to a moment of trembling or shaking the firearm prior to discharge. It typically transpires when the shooter anticipates the gunshot's recoil or commotion. One's aim may be disrupted by a downward or lateral movement caused by anticipation. This must be overcome through mental discipline and concentration on the fundamentals of trigger control and sight alignment. By focusing on your technique, you will acquire the skill of maintaining a steady aim without having to anticipate the projectile through practice.

Poor Follow-Through

Frequently, follow-through is neglected when it comes to firearm technique. It pertains to the act of preserving one's sight image and grip subsequent to the discharge of the shot. An immediate tendency among numerous shooters is to unwind or lower their arms subsequent to the release of the trigger, which can cause a disruption in the alignment of the sights. Maintain your attention on the front sight and your hold tight until you observe the impact on the target in order to increase your accuracy. By adhering to appropriate follow-through, one can enhance their ability to analyze and make adjustments to their shots.

In conclusion, Target shooting is an activity that requires concentration, accuracy, and skill. You can improve your overall shooting prowess and accuracy by avoiding these frequent errors and honing your shooting fundamentals. It is essential to bear in mind the following: maintain a firm grip, utilize proper sight alignment and picture, execute a smooth trigger draw, prevent anticipation, and ensure follow-through. You will be well on your way to developing into a more proficient and accurate shooter with diligence, consistent effort, and perseverance.

CHAPTER 5

CLEANING AND MAINTANANCE

The procedures for cleaning a handgun appropriately vary depending on whether one is performing a field stripping or a comprehensive disassembly. Fundamental field removal will be discussed in this article. The comprehensive disassembly of the slide and/or frame is not addressed in this article due to the wide range of handguns.

1. Thoroughly de-energize the handgun

This phase holds the utmost importance throughout the cleaning procedure. In order to safeguard oneself and others, it is imperative to remove any primed ammunition and the magazine from the firearm. Do so while ensuring that the firearm is pointed in a secure direction. Ensure you conduct a comprehensive visual (look) and tactile (feel) inspection. Never depend on the safety of your firearm; accidents do occur.

2: Handgun Cleaning

For semi-automatic pistols only, remove the slide from the frame in accordance with the manufacturer's take-down instructions. Be vigilant for bushings, springs, and other small components, and place them in a container such as a cup, can lid, or similar receptacle to prevent their misplacement. Using a cleaning instrument, such as a cleaning swab or utility brush, eliminate any loose particulates from the slide, cylinder (revolvers), frame, and chamber. Inspect carried handguns' interior frame regions for fibers and dust/dirt with extreme care. A thin layer of cleanser can be utilized to eliminate stubborn fouling before brushing. Clean fouling by removing it with a cloth or swab. Applying bore cleanser to a cleaning patch and passing it through the barrel with a cleaning rod equipped with a jag, brush, or patch holder tip constitutes the next step. The degree of cleanliness being achieved with a firearm can be determined by observing the quantity of detritus or residue that is removed when a patch is passed over it. Before reversing, ensure that the brush point has passed completely through the barrel to prevent it from becoming entangled. To ensure that the barrel is dry and spotless, wipe your bore cleaner-soaked patch with a series of dry patches. For optimal barrel/chamber cleanliness, swabs are useful in this situation because they can be used to clean difficult-to-reach areas of the slide and frame. In cases where barrels and compartments are severely fouled, a more robust cleaner may be employed.

3. Lubricate in areas where it is required

After thoroughly cleaning the slide, frame, barrel, and chamber of your firearm, you will need to lubricate the moving parts with the appropriate oil.

This will depend on the model of firearm you are cleaning and the manufacturer-recommended lubrication. Generally speaking, you will need to lubricate: Lubricants should only be applied to the interior of the barrel or chamber when storing it for an extended period of time It is not always the case that more lubrication is superior when it comes to firearms. Lubricate where necessary, but avoid lubricating excessively; doing so can result in excessive residues, which can cause malfunctions and handling issues (slippery). Lubricating linens are an exceptional instrument for regulating the quantity of lubrication or oil applied to a specific area of your firearm. Examine areas of deteriorated metal on bearing surfaces, including the barrel, frame rails, and slide. Certain areas of a used firearm will have exposed metal beneath the finish. This does not necessarily indicate a malfunction, but rather that your firearm is adjusting to operate more efficiently. Areas that are damaged need a gentle lubrication.

4. Cleanse Every Component

Wipe your firearm dry and clean after you have completed the lubrication and cleaning processes. This eliminates any oils, moisture, or residue that may have been overlooked in steps two and three regarding fingerprints. After wiping down all components, you can protect and sanitize your firearm with gun cloths treated with wax.

5. Reassemble and inspect for any irregularities

Once the handgun has been thoroughly cleansed and wiped down, it is time to begin reassembling it. It is imperative to conduct a thorough examination of every component while assembling it, noting any instances of impairment, irregularities, excessive wear, or play (movement). Occasional scratches or blemishes occur. Additional harm to your firearm not only diminishes its value but also raises possible safety concerns. We strongly advise you to have any suspicious items discovered during your inspection transported to a nearby firearms shop or a qualified gunsmith. They will conduct a more thorough examination and provide guidance on how to proceed in order to maintain ongoing safety and dependability.

Importance of Pistol Maintenance:

Maintaining and cleaning your pistol is equally as essential as practicing on the range. In fact, maintaining your pistol in pristine condition is just as crucial as maintaining your automobile. Primary and foremost, consistent maintenance will guarantee that your pistol remains operational at all times and prevent any potential malfunctions. There are numerous compelling justifications for regularly cleaning and maintaining your pistol.

• It is possible to guarantee that your pistol is consistently operational by performing routine maintenance, which aids in the prevention of malfunctions.

• Prolongs Lifespan: Consistent maintenance will preserve the functionality of your pistol for many years, extending its lifespan.

• Improves Reliability: Proper firearm maintenance leads to greater accuracy and reliability.

• Saves Funds: If you do regular maintenance, you'll escape having to pay a lot of money for repairs and replacements.

• Safeguards Your Investment: Maintaining a pistol appropriately is crucial in safeguarding the value of an investment in this type of firearm.

Tips on Storage

1. Safeguard Your Guns Strictly clean firearms on the inside and outside. Even plastic and copper are capable of attracting moisture. Keep your firearms dry. Likewise, clean the bore down. Apply a TINY amount of lubrication to all metal surfaces in order to form a protective coating. Wax your wood stocks on the inside and out at all times.

This will aid in the prevention of wood swelling and splitting.

2. Eliminate fingerprints from firearms

Rusting may penetrate through natural salts left on surfaces, such as fingerprints, which are transferred from the palms. When storing, wipe them with an ammonia-free cleanser.

3. Avoid overcrowding your gun safes; firearms frequently sustain nicks and dents during the process of entering and exiting a gun safe. Avoid overloading and prepare the placement in advance.

4. Restrict Humidity Levels Within Your Gun Safe. Preserve the air inside a gun safe at a consistent 50 percent relative humidity or slightly dry throughout the year. Maintain the environment by utilizing a dehumidifier and a hygrometer. Periodically inspect firearms that have been stored, and more frequently if you reside in a humid climate.

5. Bolt your gun safe down. It will be difficult for a thief to remove the safe from your residence or other location by tipping it over. Ensure that the floor can adequately bear the weight of the safe. It is possible that additional joists will be required to support the gun safe's weight and fastening mechanism.

6: Do not store your guns in gun cases. While silicone-impregnated gun socks, such as the silicone treated gun sock from Cabella, are suitable for gun safe storage, firearms should never be stored in cases made of impermeable materials, fabrics, leather, or anything non-breathable. Guns ought to be stored in an area with adequate dry air circulation.

Made in the USA
Middletown, DE
07 September 2024